-2019/5/31-

Vibration Analysis Certification
Exam Preparation Package
Certified Vibration Analyst Category I

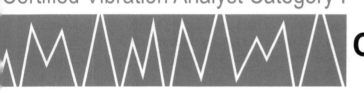 Condition Monitoring

Ali M. Al-Shurafa

Title	Vibration Analysis Certification Exam Preparation Package Certified Vibration Analyst Category I: Condition Monitoring	**Author**	Ali M. Al-Shurafa
Subtitle	ISO 18436-2 CVA Level 1: Part 4	**Series**	CAT I PREP I SERIES Practice Tests for CVA
ISBN-10	1-64415-005-0	**ISBN-13**	978-1-64415-005-4
Publisher	Prep Certify	**Ordering**	www.prepcertify.com

Copyright © 2019

All rights reserved. This intellectual asset is legally available only on a hard copy format. No part of this publication may be reproduced, translated, distributed or transmitted in any form or by any means, including photocopying, scanning, or other electronic or mechanical methods, without the prior written permission from the author except as provided by United States of America copyright law.

Disclaimers

Although every precaution has been taken to verify the accuracy of the information contained herein, the author and publisher assume no responsibility for any errors. No liability is assumed for damages that may result from the use of this package. The questions covered in this booklet are related to the standard body of knowledge according to ISO standard 18436-2 Second Edition 2015.

Comments can be sent to info@prepcertify.com

Notes

TABLE OF CONTENTS

Cat I Prep I and Your Certification 9
Cat I Part 4 BoK 15
Question Bank 17
Empty Answer Sheet 1 160
Empty Answer Sheet 2 166
Answer Key 172
Order other parts of Cat I Prep I Package 176

Notes

CAT I PREP I AND YOUR CERTIFICATION

Part 4 of Cat I Prep I Package focuses on Condition Monitoring which is a wider subject than vibration analysis. To be a specialist in condition monitoring, you need to be familiar with the commonly employed technologies and the general approaches that are practiced in the industrial maintenance. Condition monitoring is very much related to the asset management in which the maintenance strategies are selected. To develop a good common sense in this area of Specially, one would need to interact with plant assets and connect the dots between the observations collected and failure modes. Refer to Figure 1 and Table 1 for more details.

Condition Monitoring has been expanding as a maintenance strategy in the recent years thanks to the technology development of more useful and affordable hardware.

To help clarify the main maintenance strategies, refer to Figure 2 as a guide. Planning a maintenance takes place in a variety of forms and can mean different things to different people.

Understanding the concepts of Condition Monitoring helps setting up good PM and PdM programs. These programs are essential to avoid unplanned downtime in the current industry. Keep in mind that a condition monitoring program aims at two primary objectives:

1. <u>Early detection of faults</u>. If a fault is reported after the condition turns unacceptable or irreversible, the condition

monitoring program needs improvements.

2. <u>Short listing of credible maintenance actions</u>. A mature program does not give generic or long list of recommendations when a fault is detected on a machine. Recommendations such as: "take the pump to the shop for inspection and overhaul" are usually issued by condition monitoring programs that are not well administrated.

A current condition monitoring program for industrial facilities covers many technologies all of which aid the user to quantify the observations in a standard and repeatable way. Each technology in the condition monitoring program has its own scope or range of applications outside which its application is not effective. The limitations or boundaries are necessary to be recognized by (of each condition

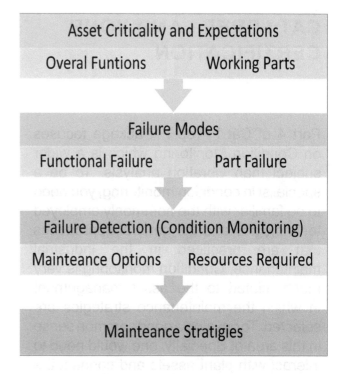

Figure 1 Maintenance Strategies Benefit from Condition Monitoring to Identify and Detect the Failures

Table 1 Examples of Condition Monitoring for Rotating Equipment

Condition Monitoring Technology	Asset Types	Faults /Damages
Vibration Thermography Motor analysis Ultrasound Oil analysis	Fans Compressors Pumps Electric motors Pressurized systems Lube oil systems	Mechanical unbalance Shaft misalignment Electrical defect (stator/rotor) Gas leakage to atmosphere Internal passing through valves Excessive heat Oil contamination Low capacity

monitoring technology) the program operator so a selection of a suitable technology can be made.

For those who are interested in deeper understanding of each technology, they may seek some dedicated training programs. Some of these trainings lead to individual certifications similar to the ISO based vibration analysis certification. Refer to Table 2 for more details.

The contents of this book are very useful for the sections covered by the Body of Knowledge in ISO 18436-2. Also, Candidates for those certification programs

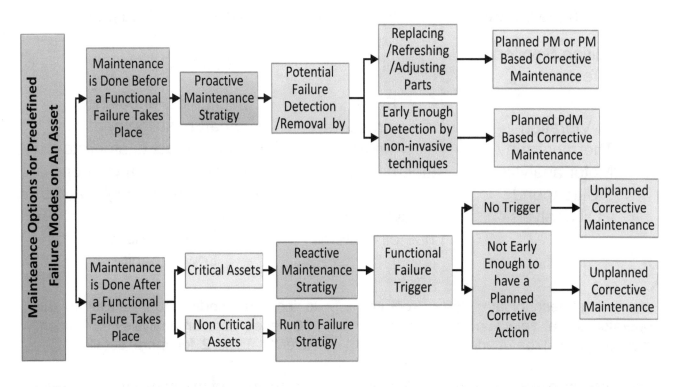

Figure 2 Maintenance Strategies in Relations with Planning and Failure Detection

(Table 2) will find this book very useful for sections associated with condition monitoring, in these certifications.

In this book, a great deal of focus is on rotating equipment, however, the concepts are possible to apply on other types of assets. The majority of the questions in the Question Bank is on pumps, electric motors, pressurized systems and lube oil systems.

Table 2 Certification Programs Covering Condition Monitoring in their BoK

	Acronym	Certification Title	Certification Body
1	CVA	Certified Vibration Analyst	Many providers
2	CMRP	Certified Maintenance and Reliability Professional	SMRP
3	CMRT	Certified Maintenance and Reliability Technician	SMRP
4	ARP	Asset Reliability Practitioner	Mobius Institute™
5	CRL	Certified Reliability Leader®	AMP

Part 4 Question Bank starts with the international standards that are commonly referred to when dealing with condition monitoring programs. Then it covers maintenance management and the common condition monitoring technologies: vibration, thermography, ultrasound and oil analysis.

The symptoms of common machinery faults are covered in many questions. Special attention is dedicated to basic trend analysis which is very useful for vibration and many other applications. Samples of spectra of basic machinery faults are also covered. Recognizing basic pre-set fault conditions, e.g. unbalance, looseness, misalignment, bearing noise and damage

As in all Cat I Prep I books, some definitions of key terms are addressed the question bank. From the 140 questions, more than 70 questions are on vibration. More than 20 questions are supported with vibration plots (trend, waveform and spectra) to analyze. A few questions require basic calculations.

Unlike to Part 5 (Fault Diagnosis), the concepts here are generic and the questions include different technologies. Most of the questions in Part 5 are geared more towards vibration signatures analysis.

Good luck- Ali M. Al-Shurafa

CAT I PART 4 BOK

Below is the Body of Knowledge as specified by the standard ISO18436-2 for Cat I Part 4 "Condition Monitoring". For a complete list of topics for Cat I, refer to Part 8 or the Standard.

#	Topic	Details
4.08	Fault condition recognition	Recognizing basic pre-set fault conditions, e.g. unbalance, looseness, misalignment, bearing noise and damage

Notes

QUESTION BANK

2019

Ali M. Al-Shurafa

Condition Monitoring
Question 001

> 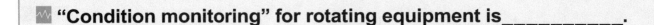 "Condition monitoring" for rotating equipment is_____.

A. watching the operating parameters through computer screens or monitors.

B. a system used in the industry to develop vibration specialists by providing a mentor or an advisor to the new employees.

C. a study conducted in laboratories and research centers to evaluate the condition of a machine against the design data.

D. applying technologies such as vibration analysis in order to have clearer and earlier assessment of the health of an asset or to flag an early detection of equipment's faults.

Condition Monitoring
Question 002

For machinery condition evaluation during acceptance testing, international standards such as ISO, NEMA, API and AGMA are commonly used. What does NEMA stand for?

A. Non-electric Manufacturing Alliance

B. Nano-electronic Machinery Association

C. Non-drive End Mechanical Applications

D. National Electrical Manufacturers Association

Note :

Condition Monitoring
Question 003

> For machinery condition evaluation during acceptance testing, international standards such as ISO, NEMA, API and AGMA are commonly used. What does ISO stand for?

A. Internal Society Organizers

B. Indiana South Organization

C. Intercontinental Standards of Origin

D. International Standardization Organization

Note :

Condition Monitoring
Question 004

For machinery condition evaluation during acceptance testing, international standards such as ISO, NEMA, API and AGMA are commonly used. What does API stand for?

A. Asset Professional Institute

B. American Petroleum Institute

C. Association of Petrochemical International

D. Automated Protection Instruments

Note :

Condition Monitoring
Question 005

> For machinery condition evaluation during acceptance testing, international standards such as BS, CE, IEEE, and AGMA are commonly used. What does AGMA stand for?

A. Association of Gear Mesh of America

B. Americana General Motor Association

C. American Gear Manufacturing Association

D. Americana Generators and Motors Associates

Note : ....

....

....

Condition Monitoring
Question 006

For machinery condition evaluation during acceptance testing, international standards such as BS, CE, IEEE, and AGMA are commonly used. What does CE stand for?

A. Certificate of Efficiency

B. Capacity Estimated

C. Conformité Européenne, meaning European Conformity

D. Certified by Engineering

Note :

Condition Monitoring
Question 007

As a vibration analyst, what abnormalities can you and should you report after conducting a field data collection survey from machines?

A. Abnormal audible noise from bearings, couplings or pipes

B. High bearing temperatures

C. Visible leakage from bearings or flanges

D. All of the above

Note : ....

....

....

Condition Monitoring
Question 008

Assume you conduct route data collections as a vibration analyst. Should you report all encountered unusual field observations and abnormalities related to the equipment being surveyed?

A. Yes, an example is an abnormally dark or leaking lube oil.

B. No, I need to focus on the specialized vibration activities.

C. No. In fact, reporting general observations is a basic operator's responsibility.

D. B and C.

Note :

Condition Monitoring
Question 009

> **What is "precision maintenance" in the modern field of industrial rotating equipment?**

A. Executing the whole planned maintenance scope with a very accurate budget estimation.

B. Maintenance activities that use microscopes during the "as-found" measurements.

C. Maintenance activities performed only by Cat III and Cat IV vibration analysts.

D. An industrial set of practices that require tighter acceptance tolerances than others practices. Examples are laser shaft alignment, rotor balancing and lube oil cleaning performed to low targets.

Question 010

 What is true about a "breakdown maintenance" strategy?

A. Owner does not perform any planned maintenance activities on the assets under this strategy.

B. Owner starts the work of the replacement/repair of an asset after it fails.

C. Strategy adapted by many owners on non-critical assets.

D. All of the above

Note :

Condition Monitoring
Question 011

What is a typical application of a breakdown maintenance strategy?

The replacement of a failed _____.

A. general-purpose lamp in a non-critical plant system.

B. general-purpose fuse in a non-critical plant system.

C. general-purpose filter in a critical plant system.

D. A and B

Note :

Condition Monitoring
Question 012

> What does an equipment owner do following a "preventive maintenance strategy" in the current industry?

A. Replace/repair an asset after it fails.

B. Execute maintenance activities (e.g. replace filters, exercise values, refresh materials etc.) to extend the functionality of an asset.

C. Prepare a plan of the maintenance tasks including the cost, man-hours, spares etc.

D. B and C

Note : ....
....
....

Condition Monitoring
Question 013

Consider a plant operating many rotating equipment. What is true about a "calendar based maintenance" in the current industry?

A. It is a planned activity whose timing starts based on a specific interval (e.g. annual or seasonal) or based on a duration (e.g. equipment's operating hours).

B. It is a maintenance strategy known also as run to failure.

C. A repair plan that avoids or prevents major activities during holidays in the company's operational calendar.

D. B and C

Note :

Condition Monitoring
Question 014

What is a disadvantage of a classic preventive maintenance strategy applied on large critical rotating equipment compared to condition based maintenance strategy?

A. PM may result in unnecessary down time.

B. PM may result in introducing new faults.

C. PM requires planning and scheduling.

D. A and B.

Note :

Condition Monitoring
Question 015

> Vibration can be measured during "Acceptance Tests" or as a part of "Predictive Maintenance" programs. What is the difference between the two cases?

	Acceptance Testing _____.	Predictive Maintenance _____.
A.	focuses on deciding if the machine can be handed over from one party to another based on the vibration readings.	focuses on machines that are in service to give operations/ maintenance staff an advice on an early fault development.
B.	is conducted online.	is conducted offline.
C.	is performed over a relatively short period of time.	is performed on a routine basis.
D.	A and C	

Condition Monitoring
Question 016

Assume you are in a facility operating many pumps and fans. Overhauling one of these pumps can be a correct preventive maintenance strategy. True or False? Explain.

A. True. A pump overhaul as a PM activity can be the best option.

B. True. A pump overhaul as a planned activity is the most commonly followed strategy in the industry.

C. False. Overhauling pumps must be a reactive strategy.

D. False. Overhauling pumps must be a run-to-failure strategy.

Note :

Condition Monitoring
Question 017

What do the following acronyms stand for when dealing with rotating equipment in the modern industry?

	PM	PdM	RPM
A.	Proactive Maintenance	Production Maintenance	Repair Maintenance
B.	Preventive Maintenance	Predictive Maintenance	Revolution per Minute
C.	Planned Maintenance	Product Data Management	Resolution per Machine
D.	Passive Maintenance	Planned Maintenance	Retroactive Planning and Management

Note :

Condition Monitoring
Question 018

 What is true about the "predictive maintenance strategy" in the current industry?

A. It initiates maintenance activities based on the asset's condition that is identified through applying vibration monitoring and/or other similar technologies.

B. It calls for conducting the required repairs within the predicted budget.

C. It is another name for "Preventive strategy".

D. A and C

Note : ....
....
....

Condition Monitoring
Question 019

Under what maintenance strategy do you classify the following activity? Replacement of an air filter at the inlet of a critical fan every year regardless of the actual filter element's condition.

A. Preventive maintenance

B. Precautious maintenance

C. Predictive maintenance

D. Synthetic maintenance

Note : ....

Condition Monitoring
Question **020**

> **What is an advantage of a predictive maintenance strategy for large critical rotating equipment?**

A. It provides early warnings of failures, damages and/or degradations.

B. It helps early planning for downtime and on-line repairs.

C. It provides short lists of causes of potential failures and/or faults.

D. All of the above

Note :

Condition Monitoring
Question 021

Most predictive maintenance activities share an advantage when applied on rotating equipment. They do not disturb the operation of the equipment as typical PdM tasks are conducted online. True or false?

A. True. For example, vibration and ultrasound produce the same results when equipment is online or offline.

B. True. Usually, operators don't need to change operating parameters or stop equipment to perform PdMs.

C. False, This advantage is true only for PM activities conducted online.

D. False. Most of the predictive maintenance activities require equipment shutdown.

Condition Monitoring
Question 022

What is a benefit from running a condition monitoring program at an industrial facility?

A. Providing early fault detection

B. Helping in early maintenance planning

C. Reducing operating parameters such as motor current, pump suction pressure and compressor discharge flow.

D. A and B

Note :

Condition Monitoring
Question 023

What is true about a "proactive maintenance strategy"?

A. It includes all maintenance activities excluding reactive types.

B. It focuses on maintenance activities that are done by management.

C. This strategy uses advanced technologies to detect asset's faults at very early stages before the faults affect the equipment's function.

D. A and C.

Note :

Condition Monitoring
Question 024

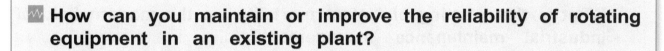

How can you maintain or improve the reliability of rotating equipment in an existing plant?

A. By applying best maintenance practices

B. By following correct operational procedures

C. By purchasing advanced vibration analysis software

D. A and B

Note :

Condition Monitoring
Question 025

 What is the meaning of "infant mortality" in the modern field of industrial maintenance management?

A. Minor defects on an asset's non-critical parts.

B. A failure type that occurs very rarely.

C. Failures that are caused by new hires with a limited experience.

D. Asset failures taking place at the early phase of an asset's life cycle.

Note : ....

Condition Monitoring
Question 026

Which of the following are common condition monitoring technologies or practices for rotating equipment in modern industry?

1. Infrared analysis (thermography)
2. Motor current analysis
3. Critical left plans for bolts, studs and nuts
4. Calculating Mean Time Between Failures (MTBF) and Mean Time to Repair (MTTR)

A. 1 and 3

B. 2 and 4

C. 1, 2 and 4

D. All of the above

Condition Monitoring
Question 027

Which of the followings are the common condition monitoring technologies for rotating equipment?

1. Vibration analysis
2. Misalignment
3. Motor grounding
4. Cathodic protection
5. Mass unbalance

A. 1 Only

B. 1, 2 and 3

C. 2, 3, 4 and 5

D. 1, 2, 3 and 5

Condition Monitoring
Question 028

Which of the following are common condition monitoring technologies for rotating equipment in the modern industry?

1. Ultrasound
2. Oil/ wear particle analysis
3. X-ray
4. Catalyst regeneration

A. 1 and 2

B. 1 and 3

C. 3 and 4

D. All of the above

Note :

Condition Monitoring
Question 029

In the modern industry, which of the following is the most reliable to detect and diagnose a rotor mechanical unbalance for a centrifugal pump driven by a horizontal motor (1500 rpm)?

A. Ultrasound

B. Oil/ wear particle analysis

C. Vibration

D. Catalyst regeneration

Note :

Condition Monitoring
Question 030

Which of the following condition monitoring technologies for rotating equipment is commonly used to detect and diagnose motor shaft misalignment?

1. Ultrasound
2. Oil/ wear particle analysis
3. Vibration
4. Motor current analysis

A. 3

B. 1 and 3

C. 1, 2 and 3

D. 1, 2, 3 and 4

Condition Monitoring
Question 031

Consider a small but critical motor (5 Hp, 1795 rpm, anti-friction bearings, grease lubricated). Which of the following condition monitoring technologies for rotating equipment is suitable to detect and diagnose a bearing damage?

1. Ultrasound
2. Oil/ wear particle analysis
3. Vibration
4. Motor current signature

A. 1 and 2

B. 1 and 3

C. 3 and 4

D. 1, 2, 3 and 4

Condition Monitoring
Question 032

Which of the following condition monitoring technologies for rotating equipment in the modern industry is common and effective to detect an insulation breakdown in a winding of a motor?

1. Ultrasound
2. Oil/ wear particle analysis
3. X-ray
4. Catalyst regeneration

A. 1 and 2

B. 1 and 3

C. 3 and 4

D. None of the above

Condition Monitoring
Question 033

For rotating equipment, what is lineup?

A. Moving driver and/or driven machines so the shafts' centerlines are collinear in horizontal and vertical directions.

B. Reducing vibration of the equipment by adding weight on the rotor at an effective location.

C. Opening/closing valves (for the driver and driven equipment) as required per the correct operational procedure.

D. All of the above.

Note : ....

....

....

Condition Monitoring
Question 034

In machinery fault diagnosis, what is "softfoot" for a rotating equipment?

A. A poor mounting condition of the vibration sensor.

B. A high flexibility condition of a coupling spacer with respect to its hub.

C. A condition of a machine's foot that introduces undesirable flexibility and/or deformations.

D. All of the above.

Note : ....
....
....

Condition Monitoring
Question 035

For rotating equipment, what is "shaft alignment"?

A. Moving driver and/or driven machines (at off condition) so the shafts' centerlines are collinear in horizontal and vertical directions (during on condition).

B. Reducing vibration of the equipment by adding weight on the rotor at an effective location.

C. Opening/closing valves (for the driver and driven equipment) as required per the correct operational procedure.

D. All of the above.

Condition Monitoring
Question 036

What is the meaning of "forcing frequency" in machinery vibration analysis?

A. The cycle at which the vibration data is collected as a part of a PM program.

B. The frequency of electric power that energizes an electric motor.

C. The frequency of the fault acting on the machine.

D. The maximum frequency seen in a vibration spectrum, known also as F_{max}.

Note :

Condition Monitoring
Question 037

What is true about a vibration "orbit"? This term is used in the condition monitoring programs.

A. Another term for shaft rotational speed, typically measured in rpm.

B. Another term for vibration survey or route.

C. Another term for torsional vibration.

D. A two dimensional plot for shaft vibration signals collected from x and y probes on the same bearing.

Note :

Condition Monitoring
Question 038

What is the difference between the amplitudes in these two vibration plot types: a frequency spectrum and an overall trend?

	Spectral Amplitudes	Trend Amplitudes
A.	Each amplitude is associated with a frequency	Each amplitude is associated with a phase angle.
B.	Each amplitude is associated with a frequency	Each amplitude is calculated from a waveform sample.
C.	The summation of amplitudes in the spectrum equals the overall amplitude.	The summation of amplitudes in the trend equals the overall amplitude.
D.	To find the amplitudes in the spectrum, a phase detector is used.	To find the amplitudes in the trend, an analyzer is used.

Condition Monitoring
Question 039

Which of the following vibration plots are common and reliable to perform diagnosis for faults like misalignment and looseness?

1. High frequency waveform
2. High resolution phase
3. Amplitude trend
4. Sideband ultrasonic

A. 1 and 2

B. 3 and 4

C. All of the above

D. None of the above

Note :

Condition Monitoring
Question 040

For rotating equipment, what is looseness?

A. Moving driver and/or driven machines so the shafts' centerlines are collinear in horizontal and vertical directions.

B. Reducing vibration of the equipment by adding weight on the rotor at an effective location.

C. Opening/closing valves (for the driver and driven equipment) as required per the correct operational procedure.

D. An undesirable condition of an equipment where the parts move relative to other parts usually due to excessive clearance.

Condition Monitoring
Question 041

Consider a simple pump driven by an electric motor. Which of the following observations is the most indicative to a bearing damage during a diagnosis?

A. Dust build up on motor's external surfaces and stator surfaces

B. Presence of water in the lube oil

C. Installed at location with a high wind speed

D. High fluid temperature inside the pump

Note : ....
....
....

Condition Monitoring
Question 042

Consider critical rotating equipment in an operating factory. Which of the following is a possible consequence (result) of running the equipment with high vibrations for a long time?

A. Machinery catastrophic failures

B. Damage to machinery parts (deformation, fracture, etc.)

C. Production slowdown or suspension

D. All of the above

Note : ......

Condition Monitoring
Question 043

Consider a simple pump driven by an electric motor. Which of the following observations is a strong indication to a rotor unbalance during a diagnosis?

1. Dust build up on motor's external surfaces.
2. Dust build up on motor's rotor surfaces.
3. Dust build up on motor's stator surfaces.
4. Dust build up on the online vibration sensor housing.

A. 2

B. 1 and 3

C. 2 and 4

D. All of the above

Condition Monitoring
Question 044

Consider a simple pump driven by an electric motor. Which of the following observations is indicative to a rotor unbalance during a diagnosis?

1. High oil level in bearing housing.
2. Dirt build up on motor's cooling fan.
3. Low oil level in bearing housing.
4. Dust build up on motor's junction box.

A. 1 and 3

B. Only 2

C. 2 and 4

D. All of the above

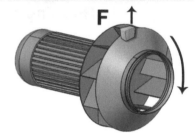

Condition Monitoring
Question 045

Consider a simple pump driven by an electric motor. Which of the following observations are reliable indications to a shaft misalignment during a diagnosis?

1. Significantly colder liquid inside a between bearing pump compared to normal operation.
2. Running the motor at 50-80% of the full load
3. Significantly hotter liquid inside a between bearing pump compared to normal operation.
4. Running the motor at 80-105% of the full load

A. 2

B. 3

C. 1 and 3

D. All of the above

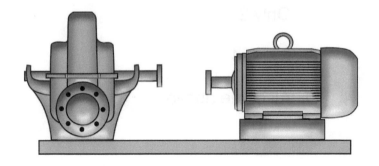

Condition Monitoring
Question 046

Consider a simple pump driven by an electric motor. Which of the following observations is are good indications to looseness during a diagnosis?

1. Missing bolts from the coupling guard.
2. Missing bolt from the name plate.
3. Excessive clearance between the impeller and case.
4. Excessive clearance between the shaft and bearing.

A. 1 and 2

B. 2 and 3

C. 3 and 4

D. Only 4

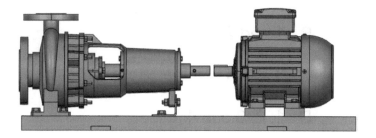

Condition Monitoring
Question 047

Consider a simple pump driven by an electric motor. Which of the following observations are important indications to looseness during a diagnosis?

1. Excessive clearance between bearing and bearing housing.
2. Excessive clearance between the impeller and shaft.
3. High oil level in bearing housing.
4. Low oil level in bearing housing.

A. 1

B. 1 and 2

C. 3 and 4

D. 1, 2, 3 and 4

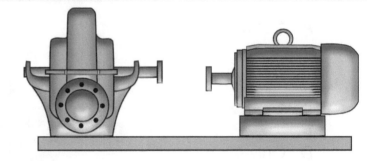

Condition Monitoring
Question 048

Consider a simple pump driven by an electric motor. Which of the following observations is the most indicative to rotor unbalance during a diagnosis?

1. Dirt build up on motor's cooling fan.
2. Loose bolts and studs on the bearing housing.
3. Broken cooling fan of a motor.
4. Low oil level in bearing housing.

A. 1 and 2

B. 2 and 3

C. 3 and 4

D. 1 and 3

Condition Monitoring
Question 049

Consider a simple pump driven by an electric motor. Which of the following observations are indicative to bearing damage during a diagnosis?

1. Low oil level in bearing housing.
2. Water in the lube oil.
3. Abnormality black lube oil.
4. Sand in the lube oil.

A. 1 and 2

B. 3 and 4

C. 2 and 4

D. All of the above

Condition Monitoring
Question 050

The following are observations collected from a horizontal motor experiencing high vibrations. What is the most possible diagnosis?
- High cable temperature.
- High amperage.
- Abnormal noise
- Hot motor case.

A. Shaft misalignment

B. Rotor unbalance

C. Electrical problem

D. Looseness

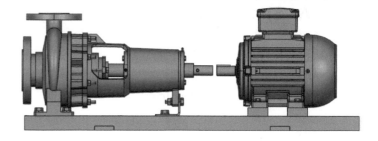

Condition Monitoring
Question 051

The following are observations collected from a horizontal pump experiencing high vibrations. What is the most possible diagnosis?

- High inboard bearing temperature.
- High service temperature inside the pump.
- Abnormal sound from the coupling.
- Dominant vibration is at 2X.

A. Shaft misalignment

B. Rotor unbalance

C. Electrical problem

D. Looseness

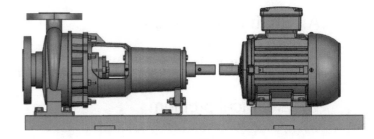

Condition Monitoring
Question 052

The following are observations collected a the vertical pump experiencing high vibration. What is the most possible diagnosis?

- **Excessive dust and grease buildup on motor's external surfaces.**
- **High 1 X amplitude on both X and Y directions.**
- **No noise or high temperature**
- **No resonance or structural problems.**

A. Shaft misalignment

B. Rotor unbalance

C. Electrical problem

D. Looseness

Note :

Condition Monitoring
Question 053

The following are observations collected from a pump with high vibrations. What is the most possible diagnosis?

- **Low oil level in bearing housing.**
- **Water in the lube oil.**
- **High bearing temperature**
- **Audible noise**

A. Shaft misalignment

B. Bearing defect

C. Electrical problem

D. Looseness

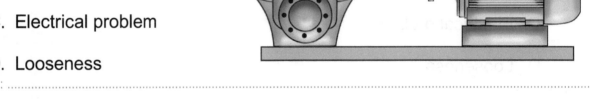

Note :

Condition Monitoring
Question 054

Which of the following problems can develop with sea water pump aging and can lead to a high vibration?

1. Pump impeller wear out
2. Case wear ring wear out
3. Shaft misalignment
4. Cracked/deformed bearing housing
5. Looseness

A. 1 and 2

B. 3 and 4

C. 1, 3 and 5

D. All of the above

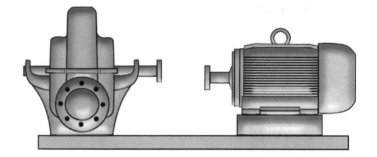

Condition Monitoring
Question 055

Which of the following problems can develop with pump aging and can lead to a high vibration?

1. Motor hold down bolt looseness
2. Grouting void development
3. Cracked/deformed bearing housing
4. Cavitation

A. 1 and 2

B. 3 and 4

C. 1, 2 and 3

D. All of the above

Note :

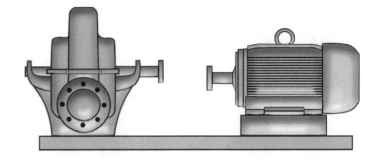

Condition Monitoring
Question 056

Which of the following problems is related to pump aging and can result in a high vibration?

1. Thermal expansion
2. High motor voltage
3. Shaft misalignment
4. Piping misalignment (suction and/or discharge)

A. 1 and 2

B. 3

C. 3 and 4

D. All of the above

Note :

Condition Monitoring
Question 057

Consider a common industrial electric motor. Which statement is true about the relation between bearing housing vibration and bearing housing temperature on rotating equipment?

A. Every time a machine malfunctions, both vibration and temperature increase.

B. As a rule of thumb, every high vibration is followed by a temperature increase.

C. As a rule of thumb, every high temperature is followed by a vibration increase.

D. Presence of high vibration and high temperature (alone or together) depends on the failure/defect type. These is no simple rule for all cases.

Condition Monitoring
Question 058

Consider a classic horizontal medium size centrifugal pump. Which of the following parts is more subject to mass unbalance if the pump's service is seawater?

A. Shaft

B. Impeller

C. Radial bearings

D. Thrust bearings

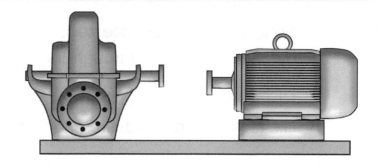

Note : ....
....
....

Condition Monitoring
Question 059

Assume you monitor a group of rotating equipment in a plant and want to have a better understanding of their problems. What sources of information should you refer to?

A. Equipment maintenance records and physical observations reported by persons walking through the plant.

B. History of oil samples, thermography images and basic operating data like temperature, pressure and flow.

C. Equipment drawings and installation procedures.

D. All of the above.

Note : ....

Condition Monitoring
Question **060**

 Consider rotating equipment under a condition monitoring program. What abnormality is difficult to detect using typical vibration analysis?

A. Low flow rate from a centrifugal pump

B. Contaminated lube oil with water in the bearing housing

C. Severe condition of cavitation on a centrifugal pump

D. A & B

Note : ....
....
....

Condition Monitoring
Question 061

What is a common fault in a typical industrial motor that can be easily detected by vibration technology?

A. Structural resonance

B. Bearing defect

C. Rotor unbalance

D. All of the above

Note :

Condition Monitoring
Question 062

Which of the following are possible causes of high vibration for rotating equipment?

A. Bad designs and manufacturing errors

B. Installation mistakes and wrong maintenance practices

C. Wrong operation and wear out of parts

D. All of the above

Note :

Condition Monitoring
Question 063

> **Consider typical vibration measurements on a faulty electric motor. What details do you need to look at when performing a basic analysis on a vibration spectrum?**

A. Highest amplitudes and their frequencies

B. Relations between key frequencies and shaft speed

C. Phase angle difference between 1X and 2X

D. A and B

Note :

Question 064

What is a key symptom of rotor unbalance?

A. The main frequency component is at 1X running speed.

B. The main frequency component is at 1 Hz for all running speeds.

C. The main frequency components are at 1X and 1/2X running speed.

D. A sudden increase in the frequency resolution.

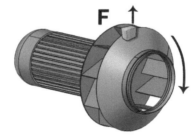

Condition Monitoring
Question 065

> **Consider typical rotating equipment. What is a key symptom of looseness?**

A. The main frequency components are at 1X, 2X, 3X and 4X running speed.

B. Generally, the vibration amplitude decreases with the speed.

C. Oil in bearing housing is dark, discolored or dirty.

D. All of the above.

Note :

Condition Monitoring
Question 066

Consider typical rotating equipment. What is a key symptom of shaft misalignment?

A. The main frequency components are at 1X and 2X running speed.

B. The main frequency components are at 1X, ½ X, 1/3 X and ¼ X running speed.

C. The main frequency component is at ½ X running speed.

D. Vibration spectrum shows noticeable non-integer orders of the shaft's running speed.

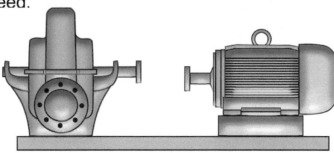

Condition Monitoring
Question 067

 What is a key symptom of a rolling element bearing damage installed in a typical pump?

A. A gradual increase in the Data Collection Time.

B. The main frequency components are at 1X, ½ X, 1/3 X and ¼ X running speed.

C. Vibration spectrum shows noticeable synchronous components. Bearings are cold or running quiet.

D. Vibration spectrum shows noticeable non-synchronous components. Bearings are hot or running noisy.

Condition Monitoring
Question 068

What is a key symptom of an anti-friction bearing damage installed in a typical pump?

A. Constant phase angle of 1x.

B. Oil in bearing housing is dark, discolored or dirty.

C. Waveform plot is rich with many frequencies.

D. B and C

Note :

Condition Monitoring
Question 069

Consider a simple fan driven by a motor. What is a key symptom of rotor unbalance?

A. The main frequency component is at 2LF.

B. Clear increase in F_{max}.

C. The main frequency components are at 1X, ½ X, 1/3 X and ¼ X running speed.

D. None of the above

Note :

Condition Monitoring
Question 070

What is the most indicative symptom of a rotor unbalance in a typical rotating equipment? Select from the given choices.

A. The main frequency component is at vane pass frequency.

B. The main frequency components are at 1X, 2X and 3X running speed.

C. The vibration amplitude increases with the speed.

D. Bearings are hot and running noisy.

Note : ....

Condition Monitoring
Question 071

Consider a classic electric motor. What is a possible symptom of an antifriction bearing damage?

A. The main frequency component is at ½ X running speed.

B. The external surface of the bearing housing is black or dirty.

C. Vibration spectrum shows noticeable non-integer orders of running speed.

D. Waveform plot is a pure sine wave.

Note : ....
....
....

Condition Monitoring
Question 072

Consider operating parameters collected from rotating equipment at a plant. What is the meaning of "deviation from the norm" when you analyze an overall vibration trend plot?

A. A significant increase or decrease compared to the general historical trend.

B. A problem that prevents the normal data collection process.

C. Converting the amplitude units from velocity to displacement (also called integration).

D. Converting the amplitude units from displacement to velocity (also called differentiation).

Note :

Condition Monitoring
Question 073

As a mechanical fault develops (gets worse) in a rotating equipment, what are the changes an analyst would see in the overall vibration trend plots?

A. An increase in the overall amplitude.

B. An increase in F_{max}.

C. An increase in the phase of the 1x component.

D. All of the above.

Condition Monitoring
Question 074

What is the best description for the vibration trend? Select from the below terms.

A. Trending up

B. Stable

C. Impulsive

D. Harmonic

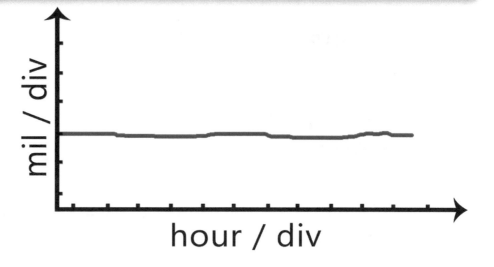

Note :

Condition Monitoring
Question 075

What is the best description for the vibration trend? Select from the below terms.

A. Trending up

B. Periodic

C. Thermo graphic parabolic

D. Pulsating

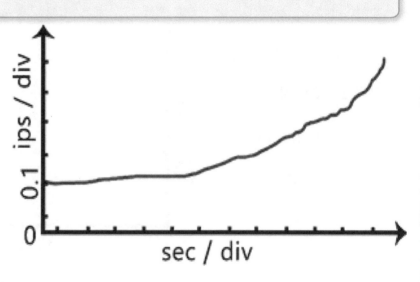

Note : ....

....

....

Condition Monitoring
Question 076

What is the best description for the vibration trend? Select from the below terms.

A. Trending down

B. High resolution

C. Sub-unity

D. Subharmonic

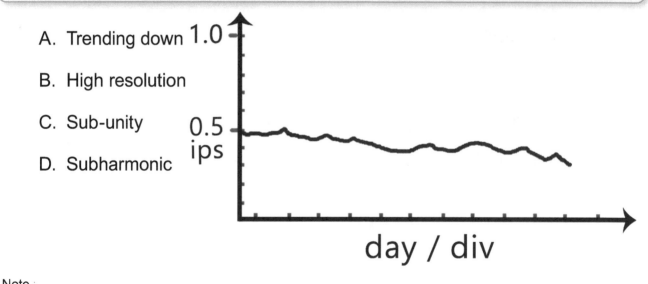

Note :

Condition Monitoring
Question 077

What is the best description for the vibration trend? Select from the below terms.

A. Low Resolution

B. Periodic

C. Pulsating

D. Ultrasonic

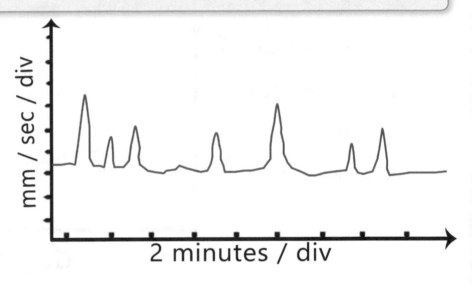

Note : ....
....
....

Condition Monitoring
Question 078

 What is the best description for the vibration trend? Select from the below terms.

A. Asynchronous

B. Periodic

C. Intermittent

D. Elastic

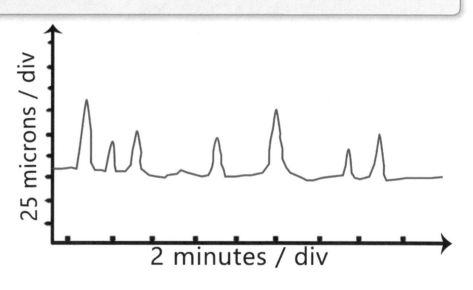

Note :

Condition Monitoring
Question 079

What is the best description for the vibration trend? Select from the below terms.

A. Periodic

B. High resolution

C. 2X Shaft speed

D. 1X Shaft speed

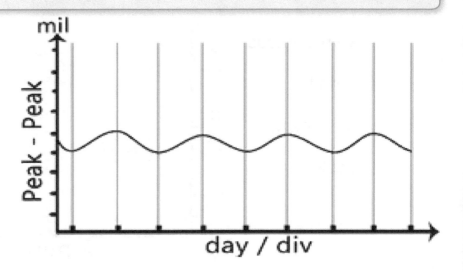

Note : ....

Condition Monitoring
Question 080

 What is the best description for the temperature trend? Select from the below terms.

A. Erratic

B. Periodic

C. Over damped

D. Thermodynamic

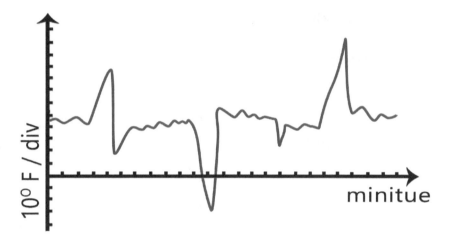

Note :

Condition Monitoring
Question 081

What is the best description for the vibration trend? Select from the below terms.

A. Periodic

B. Random

C. Sonic

D. Supersonic

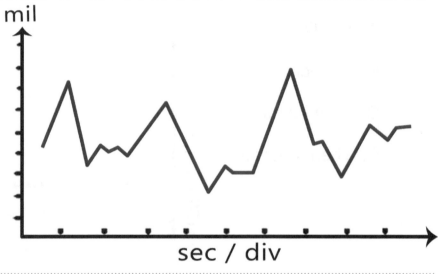

Note : ....

....

....

Condition Monitoring
Question 082

What is the best description for the vibration waveform? Select from the below terms.

A. Thermostatic

B. Periodic

C. Random

D. Elastic

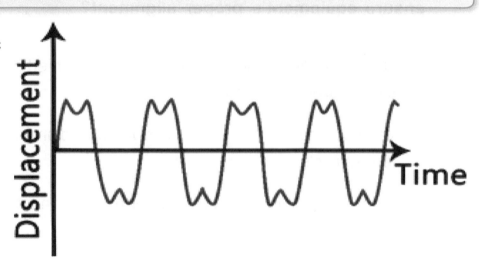

Note :

Condition Monitoring
Question 083

When installing rotating equipment, a good alignment is needed for rotors and other parts. In which direction are the shafts moved to ensure equipment's proper alignment?

A. Horizontal. Machine is moved to the right or left based on the offset and the coupling requirements.

B. Vertical. Machine is moved up or down based on the offset and the coupling requirements.

C. Axial. Machine is moved back or forth based on the offset and the coupling requirements.

D. All of the above.

Condition Monitoring
Question 084

When installing rotating equipment, a good alignment is needed. In which direction is the shaft moved to ensure equipment's proper installation?

A. Horizontal + Vertical

B. Horizontal + Axial

C. Vertical + Axial

D. Vertical + Horizontal + Axial

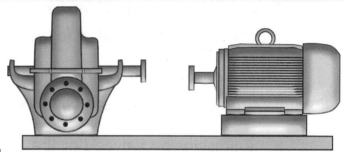

Note : ....

Condition Monitoring
Question 085

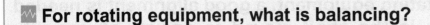 **For rotating equipment, what is balancing?**

A. Moving driver and driven machines so the shafts' centerlines are collinear in horizontal and vertical directions.

B. Reducing vibration of the equipment by adding/removing weight on/from the rotor at an effective location.

C. Opening/closing valves (for the driven equipment) as required per the correct operational procedure.

D. All of the above.

Note :

Condition Monitoring
Question 086

Consider a motor running at 2998 rpm. Which of the following spectra is the most possible signature of a rotor unbalance?

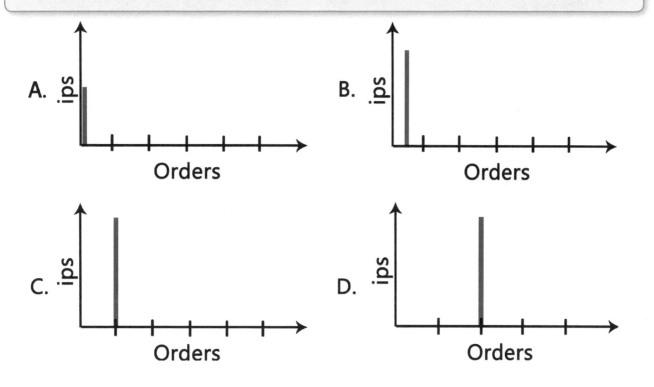

Condition Monitoring
Question 087

Consider a fan running at 225 rpm. Which of the following spectra is NOT a possible signature of a rotor unbalance?

A. 1 and 2

B. Only 3

C. 3 and 4

D. Only 4

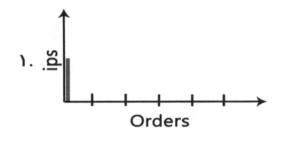

Condition Monitoring
Question 088

Consider a pump running at 995 rpm. Which of the following spectra is an indicative signature of a shaft misalignment?

A. 1 and 2

B. 3 and 4

C. Only 4

D. None of the above

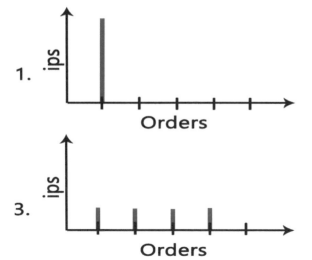

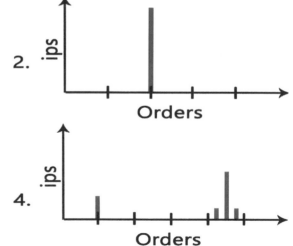

Condition Monitoring
Question 089

> Consider a four vane pump running at 995 rpm. Which of the following spectra is indicative of a shaft misalignment?

A. 1 and 2

B. 3 and 4

C. 1 and 4

D. None of the above

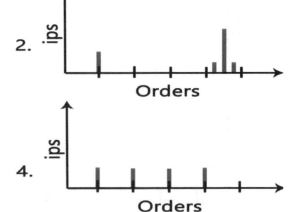

Condition Monitoring
Question 090

 Consider a pump running at 1795 rpm. Which of the following spectra is indicative of mechanical looseness?

A. 1 and 2

B. 2 and 3

C. 4

D. All of the above

Condition Monitoring
Question 091

Consider a pump mounted on antifriction bearings running at 1795 rpm. Which of the following spectra is the most possible signature of a bearings damage?

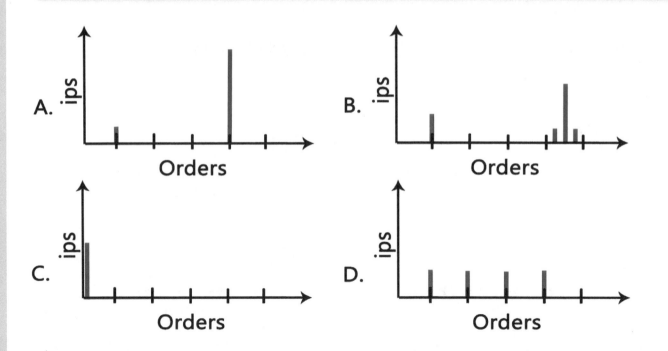

Condition Monitoring
Question 092

Consider a 4-vane pump running at 1795 rpm. Which of the following spectra is a reliable signature to confirm a poor vibration data collection? Hint : Consider a sensor mounting issue, damaged sensor, damaged cable etc.

A. A and B

B. B and C

C. C and D

D. None of the above

Condition Monitoring
Question 093

Consider a pump running at 1795 rpm. Which of the following spectra is the most possible signature of a faulty signal? Hint : Consider a sensor mounting issue, damaged sensor, damaged cable etc.

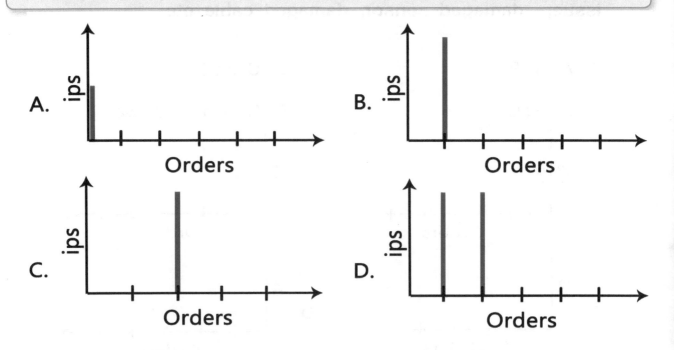

Condition Monitoring
Question 094

 What is a best description for the next vibration plot? Assume the data is from a rotating equipment.

A. Spectrum showing a possible shaft misalignment

B. Waveform showing a possible rotor unbalance

C. Orbit showing a possible a ball bearing damage

D. Trend showing a possible faulty signal from an accelerometer

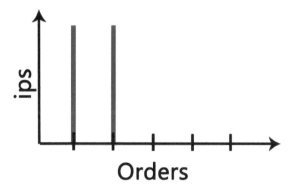

Condition Monitoring
Question 095

What is a correct description for the next vibration plot? Assume the data is from a rotating equipment.

A. Spectrum showing a possible misalignment

B. Waveform showing a possible unbalance

C. Root mean square showing signs of mechanical looseness

D. Oil analysis showing a possible fault in the data collection

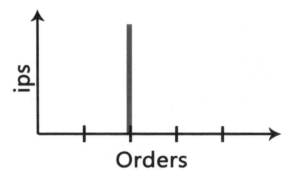

Condition Monitoring
Question 096

What is a correct description for the next vibration plot? Assume the data is from a small electric motor (10 Hp, antifriction bearing).

A. Trend showing a possible rotor unbalance

B. Waveform showing a possible antifriction bearings damage

C. Orbit showing a possible a bearing damage

D. None of the above

Condition Monitoring
Question 097

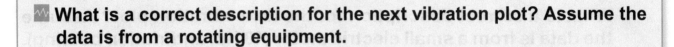

What is a correct description for the next vibration plot? Assume the data is from a rotating equipment.

A. Spectrum showing a possible looseness

B. IPS plot showing signs of uniform unbalance

C. Oil analysis showing a possible bearing defect

D. Frequency spectrum indicating and integration error

Condition Monitoring
Question 098

For plant maintenance applications, what information can be collected from the ultrasound technology applied on rotating equipment mounted on rolling element bearings?

A. Identification of very early stages of defects on bearing parts

B. Warning on lubrication problems

C. Root causes of premature coupling failures

D. A and B

Note :

Condition Monitoring
Question 099

 What is "acoustics" as a part of condition monitoring technologies applied at industrial facilities?

A. The study of sound, its transmission and its control.

B. A measure of vibration at low frequencies, typically expressed in dB.

C. A diagnostic system that uses thermal imaging to detect high temperature spots.

D. An abnormal condition of fluid flowing in pipes with a very high velocity.

Note : ....

....

....

Question 100

For plant maintenance applications, what abnormalities can be detected using ultrasound technology applied on electric equipment?

A. Arching faults

B. Contact faults in connections like grounding cables

C. Type of antifriction bearings installed

D. A and B

Note :

Condition Monitoring
Question 101

Ultrasound technology is used for many plant maintenance applications. Which of the following are common examples?

1. Estimating a pipe thickness
2. Detecting defects of rolling element bearings
3. Identifying failing steam traps
4. Balancing rotors

A. 1 and 3

B. 2 and 3

C. 1, 2 and 3

D. 1, 3 and 4

Condition Monitoring
Question 102

 Consider rotating equipment under a condition monitoring program. What abnormalities can possibly be diagnosed using a typical ultrasound device?

A. Structural resonance of a vertical motor

B. Abnormally loaded motor (e.g. 110% or 35%)

C. Faulty shaft vibration signal from an online system

D. Hot spot on a stator winding of an electric motor

Note :

Condition Monitoring
Question 103

Ultrasound technology is used for many plant maintenance applications. Which of the following is an example?

1. Identifying gas leakage from high pressure systems
2. Correcting motor unbalance
3. Estimating shaft misalignment
4. Quantifying amount of water in lube oil

A. Only 1

B. 2 and 4

C. 1, 2 and 3

D. All of the above

Note : ....

....

....

Condition Monitoring
Question 104

Ultrasound technology is used for many plant maintenance applications. Which of the following is an example?

    1. Building a temperature map of a compressor's surface
    2. Detecting defects of rolling element bearings
    3. Identifying failing steam traps
    4. Estimating clearances of rotating parts

A. 1 and 3

B. 2 and 3

C. 1, 2 and 3

D. 1, 2, 3 and 4

Note :

Condition Monitoring
Question 105

For plant maintenance applications, what information can be collected from ultrasound technology applied on gas pressurized systems?

A. Locations of sudden pressure reduction such as passing valves

B. Incorrect composition of fluid in pipes

C. Locations of gas leakage to atmosphere

D. A and C

Note :

Condition Monitoring
Question 106

 What is true about ultrasound as applied in industrial condition monitoring programs?

A. The used devices focus on sound waves with frequencies from 1 to 1000 Hz.

B. The device is of a similar size to a portable vibration analyzer but it is fixed on a specific machine.

C. The sound waves can be collected by contact and non-contact probes.

D. A and B

Note : ....

....

....

Condition Monitoring
Question 107

As a rule of thumb, what kind of a bearing lubricant is used for high-speed machines? Consider typical industrial rotating equipment.

A. Light oil

B. Heavy oil

C. Carbone grease

D. Heavy grease

Note : ....
....
....

Condition Monitoring
Question 108

Consider rotating equipment under a condition monitoring program. What abnormality can be detected using typical oil analysis?

A. Low flow rate of a lube oil pump

B. Contaminated lube oil with water

C. Severe condition of cavitation on a pump

D. Broken rotor bar of a motor

Note :

Condition Monitoring
Question 109

Consider plant maintenance applications on rotating machinery. What information can be collected from a lube oil sample analyzed in a lab?

1. Amount of water mixed with oil
2. Amount of solid dirt mixed with oil
3. Oil acidity
4. Type of thrust bearing

A. 1 and 2

B. 2 and 4

C. 1, 2 and 3

D. All of the above

Condition Monitoring
Question 110

Consider plant maintenance applications on a large pump driven by an electric motor. What information can be collected from a lube oil sample analyzed in a lab?

1. Actual oil level in the bearing housing
2. Bearing clearance
3. Type of motor's winding insulation
4. The amount of forces and moments applied on the bearings.

A. 1 and 3

B. 2 and 4

C. 1, 2 and 3

D. None of the above

Condition Monitoring
Question 111

Consider plant maintenance applications on rotating machinery. What information can be collected from a lube oil sample analyzed in a lab?

1. Viscosity
2. Oxidization level
3. Type of misalignment
4. Type of rotor unbalance

A. 1 and 2

B. 3 and 4

C. 1, 2 and 3

D. All of the above

Note :

Condition Monitoring
Question 112

Consider an oil condition monitoring program employed on rotating machinery. How can a typical laboratory analysis of an oil sample (collected from the machine) verify if the lubricant is "fit for purpose"?

A. It Helps verify the cleanliness level.

B. It Helps verify the dryness level.

C. It Helps determine the current operating temperature.

D. A and B.

Note : ....

....

....

Condition Monitoring
Question 113

Assume a high vibration was detected from a fan. Which of the following faults requires a lubrication based check as a corrective action?

A. Early stage of a bearing defect

B. Severe unbalance of fan's rotor

C. Misalignment with discolored lube oil

D. A and C

Note :

Condition Monitoring
Question 114

 For plant maintenance applications on rotating equipment, what is true about "viscosity"?

A. It is a measure of the internal resistance (of the substance) to flow.

B. It is a condition of oil when it is free from the external contaminations like sand.

C. It is a standard laboratory test used to quantify the amount and size of the contamination (usually solids) in a lube oil sample.

D. All of the above.

Note : ....

....

....

Condition Monitoring
Question 115

 For plant maintenance applications on rotating equipment, what is true about "oil cleanliness"?

A. It is a standard laboratory test used to quantify the amount and size of the contamination (usually solids) in an oil sample.

B. It is a measure of the internal resistance (of the substance) to flow.

C. It is a condition of oil describing how free it is from contaminations like sand and wear particles.

D. All of the above.

Note :

Condition Monitoring
Question 116

In the field of oil condition monitoring for rotating equipment, what is true about "particle count"?

A. It is a measure of the internal resistance (of the substance) to flow.

B. It is a standard laboratory test used to quantify the amount and size of the contamination (usually solids) in an oil sample.

C. It is a condition of oil when it is free from water and other contaminations.

D. All of the above.

Note :

Condition Monitoring
Question 117

> In the field of condition monitoring for rotating equipment, what is true about "wear particle analysis"?

A. A test done by a plant machinist on the site in order to quantify the amount of wear developed in a sleeve bearing.

B. A set of laboratory tests on lube oil samples to identify the properties of particles that are mixed with the oil due to the degradation of equipment parts (e.g. bearings and seals).

C. A standard laboratory test used to quantify the amount and size of the contamination (usually solids particles) in an oil sample.

D. B and C

Condition Monitoring
Question 118

 In wear particle analysis, which of the following is used to understand the nature of wear in a machine?

A. The material type of wear particles (e.g. metallic, polymer etc.) found in the oil sample.

B. The shape of the wear particles (e.g. round, sharp etc.) found in the oil sample.

C. The color of the wear particles (e.g. green, black etc.) found in the oil sample.

D. All of the above

Note :

Condition Monitoring
Question 119

How can a machinery analyst identify the failing components in a rotating equipment from the oil analysis results?

A. By measuring the viscosity and Total Acid Number (TAN).

B. By measuring the viscosity and Total Base Number (TBN).

C. By counting the particles of each size and putting them in standard groups.

D. By comparing the metallurgy of wear particles to those of the components in the machine.

Note :

Condition Monitoring
Question 120

What is a key difference between the wear particle tests and the tests done to evaluate the oil fitness?

	Wear particle tests _____ .	Oil fitness analysis tests _____ .
A.	focus on: contamination's quantity, size, shape, material etc.	focus on: cleanliness, dryness, oxidization etc.
B.	are conducted at the site	are usually conducted in the lab
C.	are quick and cheap to conduct	are slow and expensive to conduct
D.	All of the above	

Note : ....

Condition Monitoring
Question 121

In the modern field of industrial condition monitoring, what is "thermography/infrared"?

A. A technology (usually using portable devices like a camera) employed to measure the temperature of plant equipment.

B. A specialized lab test to detect overuse or defect development in the lube oil systems.

C. A technology applied to measure the temperature of motor's rotor bars while the machine is in service.

D. A and C

Condition Monitoring
Question 122

 Which statement is true about thermography when applied at modern plants?

A. Thermographic cameras give better results at night (e.g. at dark places).

B. Thermographic cameras give better results if the targeted surface vibrates.

C. Thermographic cameras are NOT designed for indoor applications (in rooms where lights are on)

D. A and C

Note : ....
....
....

Condition Monitoring
Question 123

For plant maintenance applications, what information can be collected from the thermography technology when applied on rotating machinery?

A. Temperatures of internal components such as motor winding temperature and pump impeller temperature.

B. Temperatures of external surfaces such as motor skin temperature and pump case temperature.

C. The amount of shaft misalignment based on the bearing temperatures.

D. A and C.

Note :

Condition Monitoring
Question 124

Consider rotating equipment under a condition monitoring program. What abnormalities can be reliably detected using typical thermography?

A. High flow rate from a pump

B. Rotor unbalance

C. Early stage of a bearing defect

D. None of the above

Note :

Condition Monitoring
Question 125

For plant maintenance applications, which of the following data can be collected by a thermography camera applied on rotating equipment?

A. Tilting pad bearing temperature

B. Bearing housing temperature

C. Gas temperature inside inlet and outlet pipes of a compressor

D. A and C

Note :

Condition Monitoring
Question 126

 In the field of industrial condition monitoring, what does IR commonly stand for?

A. Intelligent Reporting

B. Industrial Relations

C. Infrared

D. Investigation Rules

Note : ....
....
....

Condition Monitoring
Question 127

For plant maintenance applications, what information can be collected from the thermography technology applied on electric equipment?

1. Wires with abnormally high temperatures
2. Wires with abnormally low temperatures
3. Loose connections of energized cables

A. 1

B. 1 and 2

C. 3

D. All of the above

Note :

Condition Monitoring
Question 128

For plant maintenance applications, what information can be collected from the thermography technology applied on an electric motor?

1. A power cable with an abnormally excessive load on a three phase motor
2. Number of phases in a motor
3. Number of poles in a motor

A. 1

B. 2

C. 2 and 3

D. All of the above

Condition Monitoring
Question 129

For plant maintenance applications, what information can be collected from the thermography technology applied on electric equipment?

1. Phase angle of 1X, 2X and 0.5X
2. Leakage of oil from its bearing housing
3. Structural looseness such as hold down bolts.
4. Type of shaft misalignment (angular or parallel)

A. 1 and 2

B. 2 and 3

C. 3 and 4

D. None of the above

Condition Monitoring
Question 130

For plant maintenance applications, what information can be collected from the thermography technology applied on a three phase electric motor?

1. High temperature spot on the motor's case
2. Wires with abnormally low temperature
3. Loose connections of energized cables
4. A power cable with an abnormality excessive load

A. 1 and 2

B. 3 and 4

C. 1, 3 and 4

D. All of the above

Condition Monitoring
Question 131

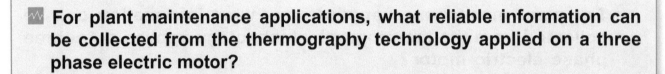

For plant maintenance applications, what reliable information can be collected from the thermography technology applied on a three phase electric motor?

A. Identification of a defective part in a rolling element bearing (inner race/outer race etc)

B. Type of lube oil contamination

C. Numbers of rotor bars and stator slots.

D. None of the above

Note : ....

....

....

Condition Monitoring
Question 132

What is "emissivity"? Hint: this term is used in relation with thermography when applied in condition monitoring programs.

A. The ability of a solid substance to absorb liquids.

B. A defect in an electric motor that can be detected by IR technology.

C. A surface property that affects the temperature read out from an IR device.

D. B and C.

Note:

Condition Monitoring
Question 133

How does emissivity affect the measurements in thermographic images?

A. The higher the emissivity configured, the higher the actual temperature of the object.

B. The higher the emissivity configured, the lower the actual temperature of the object.

C. It is a multiplication factor relating the energy measured by the IR device to the displayed reading.

D. B and C.

Note :

Condition Monitoring
Question 134

 Is there a "resolution" for thermal imaging cameras? If yes, what could that mean?

A. Yes. It is the ability of the device to distinguish between two closely separated temperature readings on the same image.

B. Yes. It is the ability of the device to provide a clearer (sharper) photographic image.

C. No. Resolution is a property that is available only in digital photographic cameras.

D. No, however, thermographic cameras operate on a similar concept called emissivity.

Note :

Condition Monitoring
Question 135

Which of the following environmental conditions could affect the thermographic images?

1. Sun light and brightness
2. Heat from direct sunlight
3. High-speed wind
4. Ambient air temperature

A. 1, 2 and 3

B. 2, 3 and 4

C. None of the above

D. All of the above

Condition Monitoring
Question 136

Which of the following could affect the reliability of the thermographic images to diagnose machinery faults?

1. Location of the measurement
2. Speed of the shaft
3. Smell from the product
4. The color and surface condition of the measurement points on the machine.

A. 1 and 2

B. 3 and 4

C. 1 and 4

D. All of the above

Condition Monitoring
Question 137

What motor abnormalities can be reliably detected using a typical motor current signature analysis?

A. Incorrect grease in bearings.

B. Excessive wear out in pump's thrust bearing.

C. Early stage of journal bearing defect.

D. None of the above.

Note : ....

Condition Monitoring
Question 138

Which of the following faults can an analyst detect by applying a motor current analysis?

A. Stator faults (e.g. winding shorts)

B. Rotor faults (e.g. broken bars)

C. Uneven air gap

D. All of the above

Note : ....
....
....

Condition Monitoring
Question 139

Consider a medium size electric motor. What is the best test to identify the condition of stator shorts and grounding faults? Select from the given options.

A. Motor current analysis

B. Lube oil analysis

C. Vibration analysis

D. Mean time between failures

Note : ....
....
....

Question 140

Assume it is the first time you grease a typical ball bearing. What would be a reasonable amount grease to be injected?

A. A size of one ball of the bearing, typically this is equivalent to 1/8 of the bearing's size

B. 10 grams per 1000 rpm

C. 25-50% of the empty volume or space between balls

D. The complete volume of the empty space between balls

Note :

Notes

Notes

EMPTY ANSWER SHEET 1

#	A	B	C	D
1	A	B	C	D
2	A	B	C	D
3	A	B	C	D
4	A	B	C	D
5	A	B	C	D
6	A	B	C	D
7	A	B	C	D
8	A	B	C	D
9	A	B	C	D
10	A	B	C	D
11	A	B	C	D
12	A	B	C	D
13	A	B	C	D
14	A	B	C	D
15	A	B	C	D

#	A	B	C	D
16	A	B	C	D
17	A	B	C	D
18	A	B	C	D
19	A	B	C	D
20	A	B	C	D
21	A	B	C	D
22	A	B	C	D
23	A	B	C	D
24	A	B	C	D
25	A	B	C	D
26	A	B	C	D
27	A	B	C	D
28	A	B	C	D
29	A	B	C	D
30	A	B	C	D

EMPTY ANSWER SHEET 1

31	A	B	C	D
32	A	B	C	D
33	A	B	C	D
34	A	B	C	D
35	A	B	C	D
36	A	B	C	D
37	A	B	C	D
38	A	B	C	D
39	A	B	C	D
40	A	B	C	D
41	A	B	C	D
42	A	B	C	D
43	A	B	C	D
44	A	B	C	D
45	A	B	C	D

46	A	B	C	D
47	A	B	C	D
48	A	B	C	D
49	A	B	C	D
50	A	B	C	D
51	A	B	C	D
52	A	B	C	D
53	A	B	C	D
54	A	B	C	D
55	A	B	C	D
56	A	B	C	D
57	A	B	C	D
58	A	B	C	D
59	A	B	C	D
60	A	B	C	D

EMPTY ANSWER SHEET 1

61	A	B	C	D
62	A	B	C	D
63	A	B	C	D
64	A	B	C	D
65	A	B	C	D
66	A	B	C	D
67	A	B	C	D
68	A	B	C	D
69	A	B	C	D
70	A	B	C	D
71	A	B	C	D
72	A	B	C	D
73	A	B	C	D
74	A	B	C	D
75	A	B	C	D

76	A	B	C	D
77	A	B	C	D
78	A	B	C	D
79	A	B	C	D
80	A	B	C	D
81	A	B	C	D
82	A	B	C	D
83	A	B	C	D
84	A	B	C	D
85	A	B	C	D
86	A	B	C	D
87	A	B	C	D
88	A	B	C	D
89	A	B	C	D
90	A	B	C	D

EMPTY ANSWER SHEET 1

91	A	B	C	D
92	A	B	C	D
93	A	B	C	D
94	A	B	C	D
95	A	B	C	D
96	A	B	C	D
97	A	B	C	D
98	A	B	C	D
99	A	B	C	D
100	A	B	C	D
101	A	B	C	D
102	A	B	C	D
103	A	B	C	D
104	A	B	C	D
105	A	B	C	D

106	A	B	C	D
107	A	B	C	D
108	A	B	C	D
109	A	B	C	D
110	A	B	C	D
111	A	B	C	D
112	A	B	C	D
113	A	B	C	D
114	A	B	C	D
115	A	B	C	D
116	A	B	C	D
117	A	B	C	D
118	A	B	C	D
119	A	B	C	D
120	A	B	C	D

EMPTY ANSWER SHEET 1

121	A	B	C	D
122	A	B	C	D
123	A	B	C	D
124	A	B	C	D
125	A	B	C	D
126	A	B	C	D
127	A	B	C	D
128	A	B	C	D
129	A	B	C	D
130	A	B	C	D
131	A	B	C	D
132	A	B	C	D
133	A	B	C	D
134	A	B	C	D
135	A	B	C	D

136	A	B	C	D
137	A	B	C	D
138	A	B	C	D
139	A	B	C	D
140	A	B	C	D

Notes

EMPTY ANSWER SHEET 2

#					#				
1	A	B	C	D	16	A	B	C	D
2	A	B	C	D	17	A	B	C	D
3	A	B	C	D	18	A	B	C	D
4	A	B	C	D	19	A	B	C	D
5	A	B	C	D	20	A	B	C	D
6	A	B	C	D	21	A	B	C	D
7	A	B	C	D	22	A	B	C	D
8	A	B	C	D	23	A	B	C	D
9	A	B	C	D	24	A	B	C	D
10	A	B	C	D	25	A	B	C	D
11	A	B	C	D	26	A	B	C	D
12	A	B	C	D	27	A	B	C	D
13	A	B	C	D	28	A	B	C	D
14	A	B	C	D	29	A	B	C	D
15	A	B	C	D	30	A	B	C	D

EMPTY ANSWER SHEET 2

#						#				
31	A	B	C	D		46	A	B	C	D
32	A	B	C	D		47	A	B	C	D
33	A	B	C	D		48	A	B	C	D
34	A	B	C	D		49	A	B	C	D
35	A	B	C	D		50	A	B	C	D
36	A	B	C	D		51	A	B	C	D
37	A	B	C	D		52	A	B	C	D
38	A	B	C	D		53	A	B	C	D
39	A	B	C	D		54	A	B	C	D
40	A	B	C	D		55	A	B	C	D
41	A	B	C	D		56	A	B	C	D
42	A	B	C	D		57	A	B	C	D
43	A	B	C	D		58	A	B	C	D
44	A	B	C	D		59	A	B	C	D
45	A	B	C	D		60	A	B	C	D

EMPTY ANSWER SHEET 2

61	A	B	C	D
62	A	B	C	D
63	A	B	C	D
64	A	B	C	D
65	A	B	C	D
66	A	B	C	D
67	A	B	C	D
68	A	B	C	D
69	A	B	C	D
70	A	B	C	D
71	A	B	C	D
72	A	B	C	D
73	A	B	C	D
74	A	B	C	D
75	A	B	C	D

76	A	B	C	D
77	A	B	C	D
78	A	B	C	D
79	A	B	C	D
80	A	B	C	D
81	A	B	C	D
82	A	B	C	D
83	A	B	C	D
84	A	B	C	D
85	A	B	C	D
86	A	B	C	D
87	A	B	C	D
88	A	B	C	D
89	A	B	C	D
90	A	B	C	D

EMPTY ANSWER SHEET 2

91	A	B	C	D
92	A	B	C	D
93	A	B	C	D
94	A	B	C	D
95	A	B	C	D
96	A	B	C	D
97	A	B	C	D
98	A	B	C	D
99	A	B	C	D
100	A	B	C	D
101	A	B	C	D
102	A	B	C	D
103	A	B	C	D
104	A	B	C	D
105	A	B	C	D

106	A	B	C	D
107	A	B	C	D
108	A	B	C	D
109	A	B	C	D
110	A	B	C	D
111	A	B	C	D
112	A	B	C	D
113	A	B	C	D
114	A	B	C	D
115	A	B	C	D
116	A	B	C	D
117	A	B	C	D
118	A	B	C	D
119	A	B	C	D
120	A	B	C	D

EMPTY ANSWER SHEET 2

#	A	B	C	D
121	A	B	C	D
122	A	B	C	D
123	A	B	C	D
124	A	B	C	D
125	A	B	C	D
126	A	B	C	D
127	A	B	C	D
128	A	B	C	D
129	A	B	C	D
130	A	B	C	D
131	A	B	C	D
132	A	B	C	D
133	A	B	C	D
134	A	B	C	D
135	A	B	C	D

#	A	B	C	D
136	A	B	C	D
137	A	B	C	D
138	A	B	C	D
139	A	B	C	D
140	A	B	C	D

Notes

ANSWER KEY

Q. #	Answer
1	D
2	D
3	D
4	B
5	C
6	C
7	D
8	A
9	D
10	D
11	D
12	D
13	A
14	D
15	D

Q. #	Answer
16	A
17	B
18	A
19	A
20	D
21	B
22	D
23	D
24	D
25	D
26	C
27	A
28	A
29	C
30	A

Q. #	Answer
31	B
32	D
33	C
34	C
35	A
36	C
37	D
38	B
39	D
40	D
41	B
42	D
43	A
44	B
45	C

Q. #	Answer
46	D
47	B
48	D
49	D
50	C
51	A
52	B
53	B
54	D
55	C
56	C
57	D
58	B
59	D
60	D

ANSWER KEY

Q. #	Answer
61	D
62	D
63	D
64	A
65	A
66	A
67	D
68	D
69	D
70	C
71	C
72	A
73	A
74	B
75	A

Q. #	Answer
76	A
77	C
78	C
79	A
80	A
81	B
82	B
83	D
84	D
85	B
86	C
87	D
88	A
89	D
90	C

Q. #	Answer
91	B
92	D
93	A
94	A
95	A
96	D
97	A
98	D
99	A
100	D
101	C
102	B
103	A
104	B
105	D

Q. #	Answer
106	C
107	A
108	B
109	C
110	D
111	A
112	D
113	D
114	A
115	C
116	B
117	B
118	D
119	D
120	A

ANSWER KEY

Q. #	Answer
121	A
122	A
123	B
124	D
125	B
126	C
127	D
128	A
129	D
130	D
131	D
132	C
133	C
134	A
135	D

Q. #	Answer
136	C
137	D
138	D
139	A
140	C

Notes

ORDER OTHER PARTS OF CAT I PREP I PACKAGE

Don't guess where your skill stands; certify it. PrepCertify believes that the best preparation for professional certifications is obtained through practicing well-designed real world problems.

Learn what really matters in real world industry while mastering the Body of Knowledge in the certification standards. Your Cat I Prep I series does that for you. Through PrepCertify, you will achieve your certification in a much shorter time and with a greater value of your time and effort.

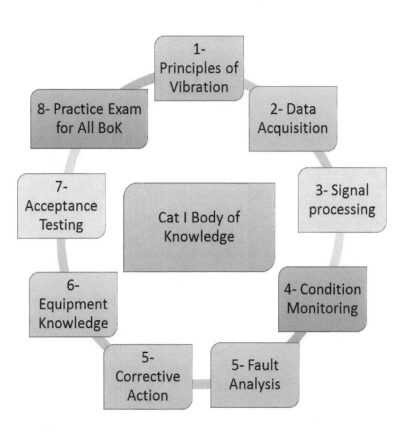

Vibration Analysis Certification Exam Preparation Package
Certified Vibration Analyst Category I
ISO 18436-2 CVA Level 1
CAT I PREP I SERIES PRACTICE TESTS

Part	Covered Body of Knowledge	ISBN
1	PRINCIPLES OF VIBRATION	978-1-64415-006-1
2	DATA ACQUISITION	978-1-64415-009-2
3	SIGNAL PROCESSING	978-1-64415-002-3
4	CONDITION MONITORING	978-1-64415-005-4
5	FAULT ANALYSIS	978-1-64415-008-5
6	EQUIPMENT KNOWLEDGE	978-1-64415-001-6
7	ACCEPTANCE TESTING	978-1-64415-004-7
8	TWO PRACTICE TESTS	978-1-64415-007-8

Notes

Notes

Notes

Made in United States
Troutdale, OR
10/09/2024